Revised & Expanded

BEND & STRETCH

Suppleness and Strength Exercises

by Gordon W. Stewart, M.Sc. and Robert A. Faulkner, Ph.D.
drawings by Chelle

3S® Fitness Group Ltd.

ISBN 0-920846-03-3
Second Edition (previously ISBN-00-9)

Cover illustration: Chelle
Cover colour: Joan Tiernan
Typesetting, design and layout: Dynagraphics
Printed in Canada by Fleming Printing Limited

Acknowledgement: Thanks to Nancy Taylor for her early enthusiasm and hard work which helped the BEND & STRETCH idea grow. And a special thanks to Arlis McQuarrie (B.A., BPT) for her assistance with this revised edition of BEND & STRETCH. Arlis is a clinical lecturer in the School of Physical Therapy, College of Medicine, University of Saskatchewan. She served as a therapist with the Canadian team at the 1983 World University Games.

PRINTED ON RECYCLED PAPER

3S® Fitness Group Ltd.

Box 5520, Station B
Victoria, British Columbia
V8R 6S4

INDEX

1 INTRODUCTION

Stamina, suppleness, and strength. These are the S's of fitness. Add a fourth item — W for weight control — and you have four important components of fitness. Straightforward and simplified.

BEND & STRETCH will help you develop suppleness and strength. Combine these exercises with aerobic activities and you have a total fitness program.

The exercise routines included in Chapter 3 offer a gentle and progressive approach to suppleness and strength improvement. Three sets of 15 exercises are provided, with each set a little more difficult than the one before.

Used as a general program, BEND & STRETCH exercises can bring a proper *balance* of strength and suppleness, thereby improving posture and reducing the risk of low back pain. To complement the Chapter 3 exercises, specific tips for back care are provided in Chapter 4.

As a warmup routine, BEND & STRETCH exercises are meant to prepare the body for action: a proper warmup allows greater ease and freedom of movement and reduces the risk of injury during vigorous activity.

A good level of suppleness and strength also aids in skill development. Athletes and fitness enthusiasts should, therefore, use the exercises as one part of an overall training program. To this end, Chapter 5 can help. It notes the muscles used and most important exercises for a variety of sports.

Before beginning the exercises, read the next chapter carefully. It provides more details on the importance of strength and suppleness; it discusses several key principles of exercise; and it notes some exercise precautions.

If you have been inactive for some time, complete PAR-Q — the brief activity questionnaire — on page 7. Examine it carefully. Answer the questions and follow the "yes" or "no" advice, whichever applies to you.

Finally, remember that suppleness and strength exercises alone do not constitute a complete program. Stamina is the first and most important 'S' of fitness. Cycle, swim, walk, hike, run, play sports. Stay active, stay healthy and fit.

2 BEFORE YOU BEGIN

Background Details

Strong and supple muscles help us accomplish our daily tasks with ease and efficiency. Supple muscles allow us to bend, turn, and twist comfortably, and strong muscles allow us to lift and carry things with less effort or fatigue.

Muscles also serve as the first line of defence against injury to the various joints of the body. Strong muscles help stabilize joints and protect the ligaments and tendons, while adequate suppleness allows the muscles to exert their forces through the full range of motion. A regular exercise routine, then, has some very obvious and important benefits.

As a warmup before fitness and recreational activities, a gentle series of suppleness and strength exercises aids in both preparation and injury prevention. The warmup helps raise heart rate, increase muscle temperature, stretch tight connective tissue at the ends of the muscles, and forces lubricating fluid into the joints. All this helps the muscles function more effectively.

Exercises play an equally important role in the cooldown following vigorous activity. A cooldown brings heart rate and body temperature back to normal, and it helps remove the waste products of muscular activity. A proper cooldown also helps prevent unnecessary stiffness and soreness.

General Principles

Do stretching exercises slowly, gently, and without pain.

When a stretching movement is done slowly, the muscle relaxes and *gives* with the movement. A fast, bouncing, or jerking movement has quite the opposite effect. In a fast movement, the muscle senses it may be overstretched and it invokes a protective mechanism known as the stretch reflex. This reflex actually causes the muscle to contract to slow the movement in an attempt to prevent injury. It is, thus, counterproductive to the development of suppleness.

Pain is the muscle's built-in indicator that overstretch is occurring. A stretch-and-hold or gentle movement approach to the exercises allows you to heed the warning sign. Fast bouncing exercises may take you beyond the pain level before you can stop, resulting in overstretch and possibly minor injury.

Do strengthening exercises at a controlled speed.

Strengthening exercises should also be done at a slow, controlled speed. This will allow you to concentrate on proper technique, thus ensuring maximum benefit and minimum injury risk.

Breathe rhythmically.

Breath holding during exertion increases the pressure within the chest cavity and puts greater demands on the heart. This can cause dizziness or faintness and may be particularly dangerous for individuals with heart problems or high blood pressure.

Exercise should be accompanied by controlled, rhythmical breathing, with an inhalation and exhalation on every repetition. For stretching exercises, exhale as you bend or twist. For strengthening exercises, exhale on effort.

Warmup and cooldown leisurely.

A good warmup consists of gentle activity *and* preparation exercises. To get the most from your suppleness and strength routine, precede it with five minutes of gentle movement (brisk walking, marching on-the-spot, etc.) For the cooldown, return to exercises that stretch the muscles which were most used during your activity. Finally, as a general guide: the more demanding the main activity and/or the less fit the individual, the more time that should be devoted to warmup and cooldown activities.

Strive for safety.
Once you've worked with the Chapter 3 routines for a while, you may want to vary your program and include some other exercises. If you do, be cautious and conservative. Avoid exercises which include straight double-leg lifts in a supine (on the back) position or exercises which hyperextend the back. These put excessive pressure on the lower back and can cause injury. In addition, don't go beyond a 90 degree bend of the knees in any squatting motion.

Be patient.
It takes time *and* regular, moderate activity to improve suppleness and strength. Be patient and persistent. Look for improvement over a period of time, not overnight.

Exercise Specifics

Learn proper technique.
Spend a little extra time your first few sessions with each of the series of exercises in Chapter 3. Read the descriptions and examine the illustrations carefully. Use a mirror or work with a partner to make sure you assume the proper body position and execute the correct movement.

If an exercise hurts or feels wrong, seek help. Have a fitness specialist check your technique. If any exercise is *still* uncomfortable, eliminate it from your program for now.

In addition to learning proper technique, you might want to consider which muscles are being conditioned in the different exercises. If so, match up the "muscles used" (as noted at the end of each exercise description) with the muscle groups as labelled on the following "muscle map."

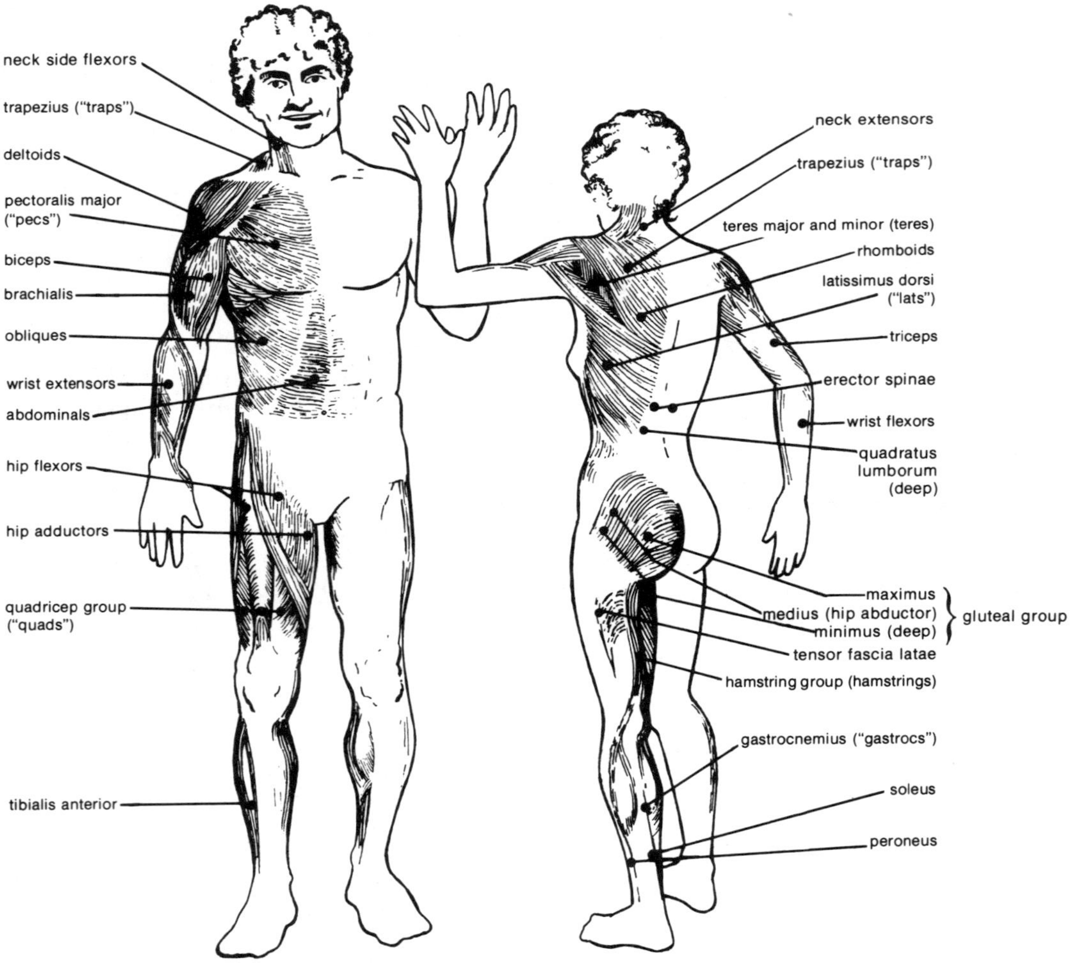

— "Muscle Map" illustration by Cheryl Campbell.

Look after your back.

The pelvic tilt exercise (#2, page 22) is of particular importance in proper back care. Try it out and get a feel for it. Its purpose is to increase flexibility of the muscles in the lower back and to give an awareness of the flat-back position so important for proper posture and for safe execution of all exercises. More details on back care follow on page 21.

PAR - Q and YOU

PHYSICAL ACTIVITY READINESS QUESTIONNAIRE (PAR-Q)*

A SELF-ADMINISTERED QUESTIONNAIRE FOR ADULTS

PAR-Q is designed to help you help yourself. Many health benefits are associated with regular exercise, and the completion of PAR-Q is a sensible first step to take if you are planning to increase the amount of physical activity in your life.

For most people physical activity should not pose any problem or hazard. PAR-Q has been designed to identify the small number of adults for whom physical activity might be inappropriate or those who should have medical advice concerning the type of activity most suitable for them.

Common sense is your best guide in answering these few questions. Please read them carefully and check the ☑ YES or NO opposite the question if it applies to you.

YES	NO		
☐	☐	1.	Has your doctor ever said you have heart trouble?
☐	☐	2.	Do you frequently have pains in your heart and chest?
☐	☐	3.	Do you often feel faint or have spells of severe dizziness?
☐	☐	4.	Has a doctor ever said your blood pressure was too high?
☐	☐	5.	Has your doctor ever told you that you have a bone or joint problem such as arthritis that has been aggravated by exercise, or might be made worse with exercise?
☐	☐	6.	Is there a good physical reason not mentioned here why you should not follow an activity program even if you wanted to?
☐	☐	7.	Are you over age 65 and not accustomed to vigorous exercise?

IF YOU ANSWERED

YES TO ONE OR MORE QUESTIONS

If you have not recently done so, consult with your personal physician by telephone or in person BEFORE increasing your physical activity and/or taking a fitness test. Tell him what questions you answered YES on PAR-Q, or show him your copy.

PROGRAMS

After medical evaluation, seek advice from your physician as to your suitability for:

- unrestricted physical activity, probably on a gradually increasing basis.
- restricted or supervised activity to meet your specific needs, at least on an initial basis. Check in your community for special programs or services.

NO TO ALL QUESTIONS

If you answered PAR-Q accurately, you have reasonable assurance of your present suitability for:

- A GRADUATED EXERCISE PROGRAM- A gradual increase in proper exercise promotes good fitness development while minimizing or eliminating discomfort.
- AN EXERCISE TEST - Simple tests of fitness (such as the Canadian Home Fitness Test) or more complex types may be undertaken if you so desire.

POSTPONE

If you have a temporary minor illness, such as a common cold.

* Developed by the British Columbia Ministry of Health. Conceptualized and critiqued by the Multidisciplinary Advisory Board on Exercise (MABE). Translation, reproduction and use in its entirety is encouraged. Modifications by written permission only. Not to be used for commercial advertising in order to solicit business from the public.
Reference: PAR-Q Validation Report, British Columbia Ministry of Health, May 1978.

* Produced by the British Columbia Ministry of Health and the Department of National Health & Welfare.

3 THE EXERCISE ROUTINES

Rules

- The exercises are arranged in a progressive manner. Do Series A your first 10 sessions, progress to Series B for 10 sessions, then Series C for 10.
- After 30 sessions, you might stay with your favourite series or pick and choose from all three. If you pick and choose, make sure you do exercises for the arms, shoulders, upper torso, lower back, stomach, hips, and legs. This ensures a total body workout.
- The stretching exercises are either a stretch-and-hold or gentle continuous movement. (Each exercise description notes the proper approach.) On the stretch-and-hold type, start with *four* or *five* repetitions, holding each repetition for *seven* or *eight* seconds. Later on, as you progress, you may actually do fewer repetitions but hold each one longer. (Fifteen or 20 second 'holds', for example.)
- There are 15 exercises in each set. Spend about 15 minutes. Experiment and find the right number of repetitions for you.
- To begin, do the exercises every other day. Later on, make it a daily routine. If you are regularly active, do the exercises each time you work out.

Reminders

- To get the most from your exercises, precede them with a few minutes of brisk walking or marching on the spot.
- Do the exercises slowly and smoothly with no bouncing or jerking.
- Stretch to the point of tightness — strive for a stretched/relaxed feeling. Never stretch into pain. Pain means you're overstretching and may be doing more harm than good.
- No breath holding! Inhale and exhale on each exercise repetition. On the suppleness exercises, exhale as you bend or twist. On the strength exercises, exhale on effort.
- Be patient. Progress slowly and don't force improvement.

Cautions

- If you have a history of minor back problems, you might begin with the exercise routine in Chapter 4 and move on to the exercises in this chapter later. When you do get to them, proceed gently with exercises A-3, A-4, B-4, B-6, C-4, and C-5.
- IF YOU ARE PRESENTLY UNDER PROFESSIONAL CARE FOR A KNOWN BACK PROBLEM, work only on those exercises you have been instructed to perform. Later, this more general routine may be appropriate.

SERIES A

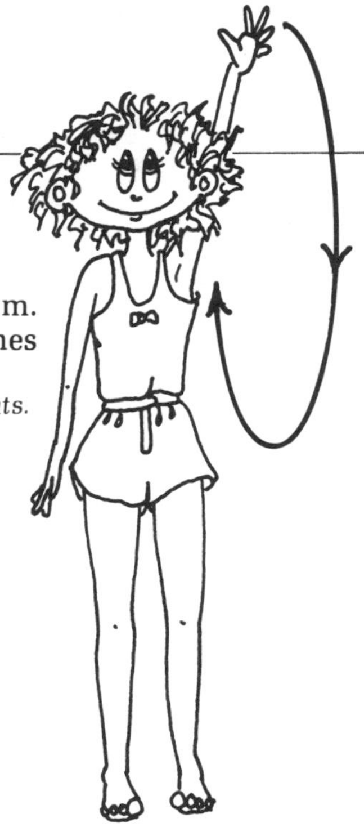

A-1 Single Arm Circle

Full, slow, sweeping circles with one arm. Forward, then backward. Exhale as arm comes down. Repeat with other arm.

Muscles exercised: pecs, deltoids, rhomboids, teres, lats.

A-2 Side Stretch

Reach one arm down the outside of the leg while exhaling. Hold. Repeat alternately to the other side.

quadratus lumborum, erector spinae, obliques.

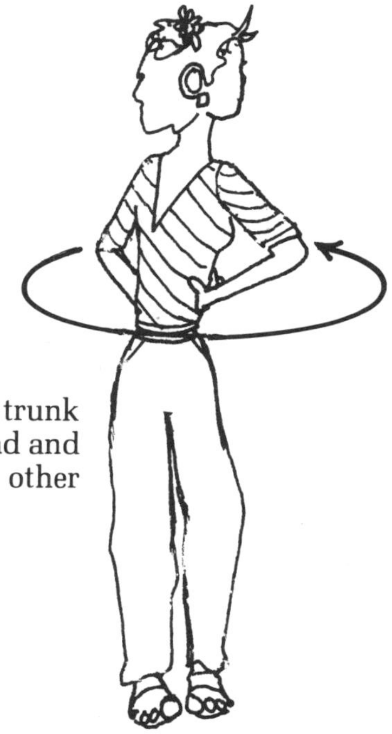

A-3 Trunk Twist

With knees slightly bent and hands on hips, trunk twist slowly in one direction, looking behind and exhaling. Hold. Repeat alternately to the other side.

erector spinae, abdominals (including obliques)

A-4 Sit Reach

Sit with one leg bent, sole of the foot near knee of the straight leg. Gently curl upper body toward knee of straight leg and reach forward with hands while exhaling. Hold. Do not bounce. Relax, then repeat. Change leg positions and repeat action.

Stretches hamstrings and erector spinae (plus gastrocs if toes are flexed).

A-5 Single Knee Press

On back, grasp hands behind one knee and bring knee toward chest while exhaling. Hold. Return to starting position. Repeat alternately with other leg.

Stretches gluteals.

A-6 Seat Lifter

Sit with legs in front, knees slightly bent, palms on the floor beside hips. Roll from one side to the other, lifting one then the other hip off the floor. (More difficult with arms folded across chest.)

Strengthens hip abductors and gluteals.

A-7 Mad Cat

On all fours, arch, tucking chin to chest and exhaling. Hold, then relax, returning to *flat back* position.

Strengthens abdominals and gluts; stretches erector spinae.

A-8 Pelvic Tilt

On your back, knees bent, feet flat on floor. Tighten abdominals and buttocks and press your lower back firmly against the floor while exhaling. Hold, relax, and repeat.

Stretches erector spinae.

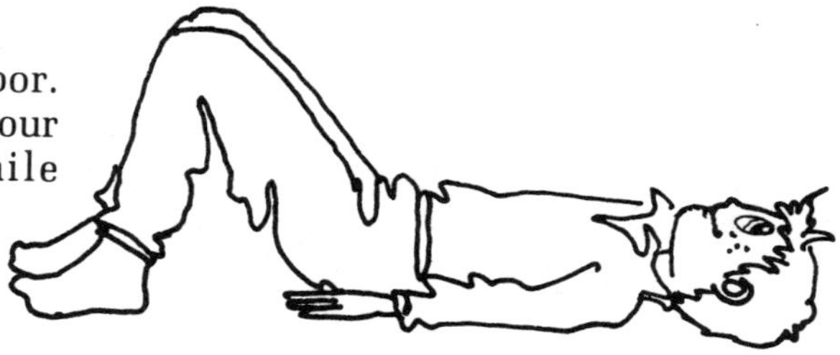

A-9 Knee Crossover

Seated, legs in front, knees bent, feet flat on floor. Roll legs to one side so the outside of one knee and the inside of the other reach toward the floor. Look over the other shoulder. Hold. Repeat alternately to the other side.

Stretches hip abductors.

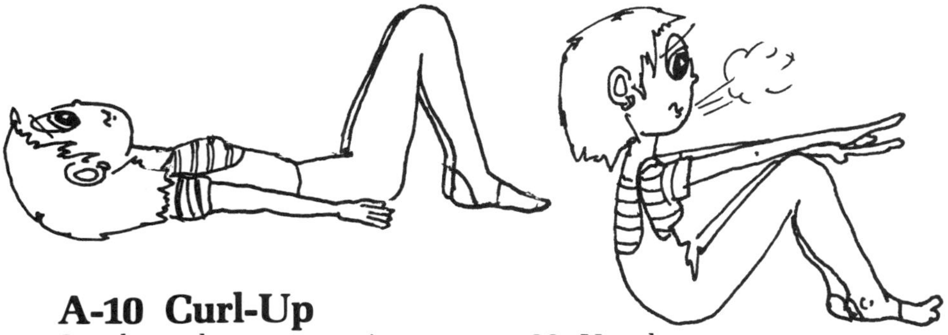

A-10 Curl-Up

See the curl-up progression on page 26. Use the technique that's right for you *now*. Move to more difficult positions as you progress.

Strengthens abdominals (including obliques if done with a twist).

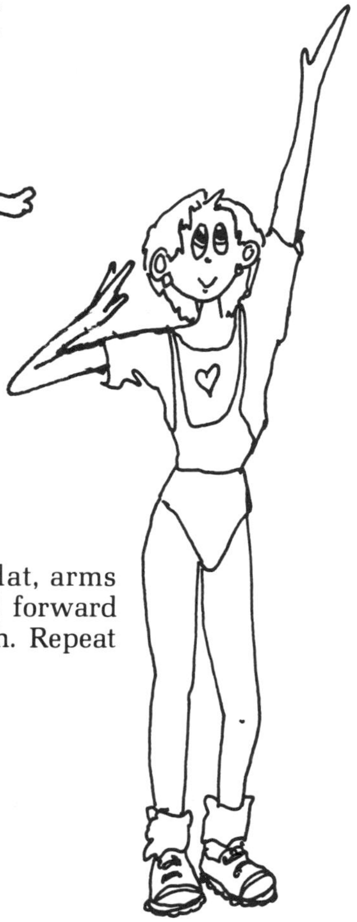

A-11 Apple Pick

Standing, knees slightly bent, back flat, arms overhead and in front. Reach up and forward with one arm, exhaling as you stretch. Repeat alternately with the other arm.

Stretches abdominals.

A-12 Thigh Stretch

Bend one knee, grasp ankle behind, pull foot gently toward seat. Hold. Relax. Repeat alternately with other leg. (Maintain a proper pelvic tilt throughout.)

Stretches quads.

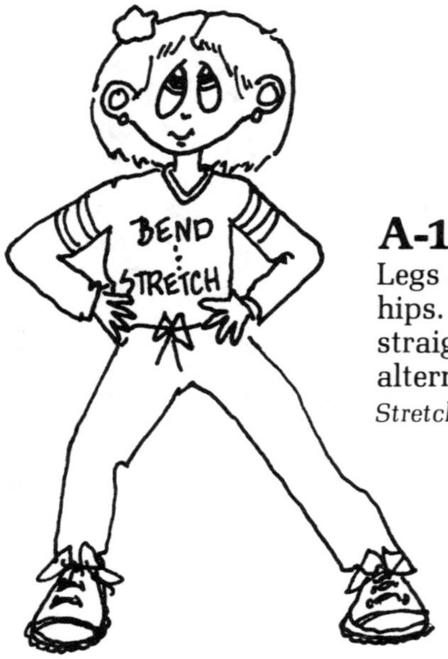

A-13 Split Stretch

Legs apart, feet pointing straight ahead, hands on hips. Shift weight over one leg, other leg kept straight and both feet *flat* on floor. Hold. Repeat alternately to other side.

Stretches hip adductors.

A-14 Soleus Stretch

Feet pointing straight ahead and one foot in front of the other, bend both legs (squatting), keeping the heels on the floor. Hold. Relax, then repeat. Change leg positions and repeat.

Stretches soleus.

A-15 'Gastroc' Stretch

Feet pointing straight ahead, one foot in front of the other, rear leg straight, front leg bent. Rock forward over front leg to stretch rear calf. Keep heel of rear foot flat on floor. Hold — don't bounce. Change leg positions and repeat.

Stretches gastrocs.

SERIES B

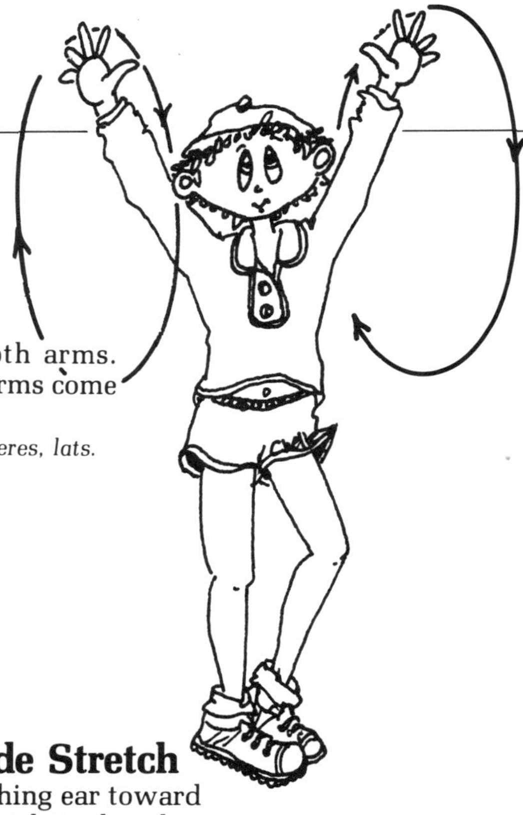

B-1 Double Arm Circle

Full, slow, sweeping circles with both arms. Frontward, then backward. Exhale as arms come down.

Muscles exercised: pecs, deltoids, rhomboids, teres, lats.

B-2 Neck Stretch & Side Stretch

Slowly tilt head to the side, reaching ear toward the shoulder. Hold. Repeat alternately to the other side and to the front (reaching chin toward chest).

Stretches neck extensors and neck side flexors.

Follow with **Side Stretch** — One hand resting on top of head, reach the other arm down the outside of leg while exhaling. Hold. Repeat alternately to the other side.

Stretches quadratus lumborum, erector spinae and obliques.

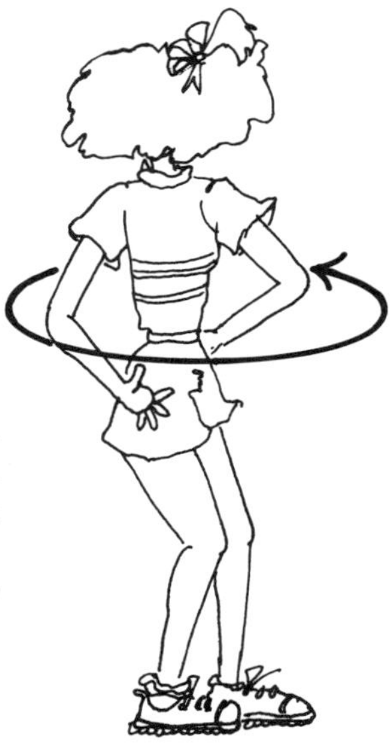

B-3 Trunk Twist

Twist trunk gently to one side, reaching the arm behind and looking behind, bending the other arm across the stomach while exhaling. Hold. Repeat alternately to the other side.

Stretches erector spinae and abdominals (including obliques).

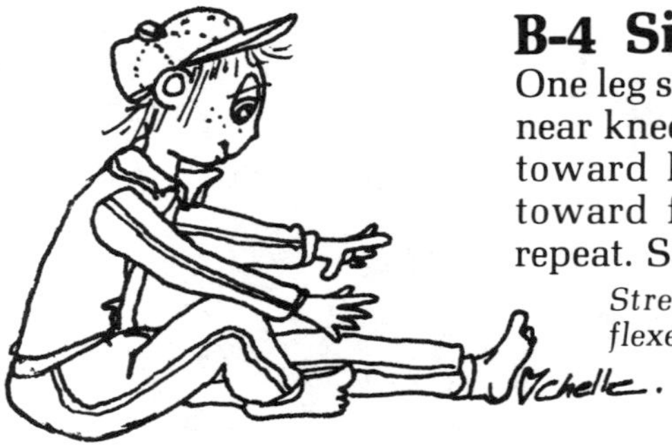

B-4 Sit Reach

One leg straight, one leg bent with sole of the foot near knee of straight leg. Gently curl upper body toward knee of straight leg and reach hands toward foot while exhaling. Hold. Relax and repeat. Same with other leg. Do not bounce.

Stretches hamstrings, erector spinae, (gastrocs if toes flexed).

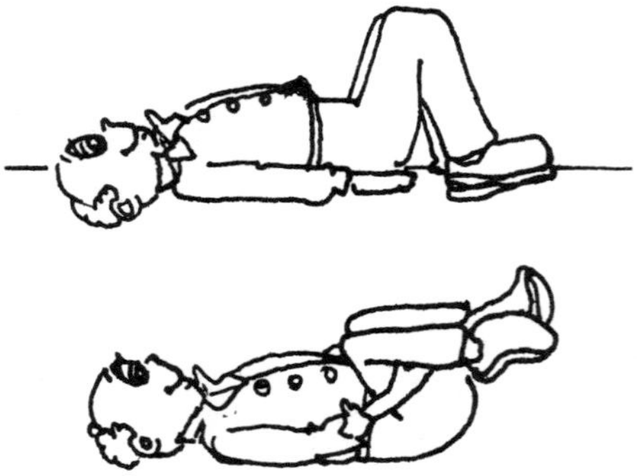

B-5 Double Knee Press

On back, legs bent, feet flat on the floor, and back pressed against floor (i.e. pelvic tilt). Pull knees to chest grasping backs of thighs with hands while exhaling. Return to starting position and repeat.

Strengthens abdominals and hip flexors, stretches gluteals.

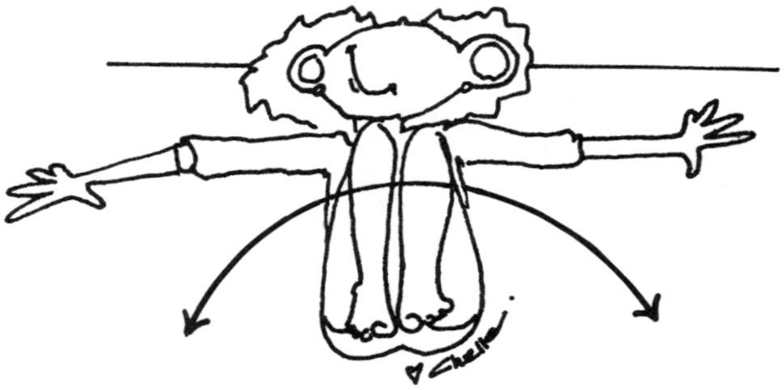

B-6 Hip Twist

On back, back pressed against floor, legs bent, arms outstretched at shoulder level, back of the head touching the floor. Draw knees toward chest then roll the legs to one side toward the floor while exhaling. Repeat alternately to the other side.

Strengthens abdominals (including obliques).

B-7 Curl & Stretch

On all fours, curl one knee in toward nose while exhaling, then slowly straighten leg behind *(parallel to the floor)*. Alternately curl and stretch. Repeat with other leg.

Strengthens gluteals and hamstrings; stretches hip flexors and erector spinae.

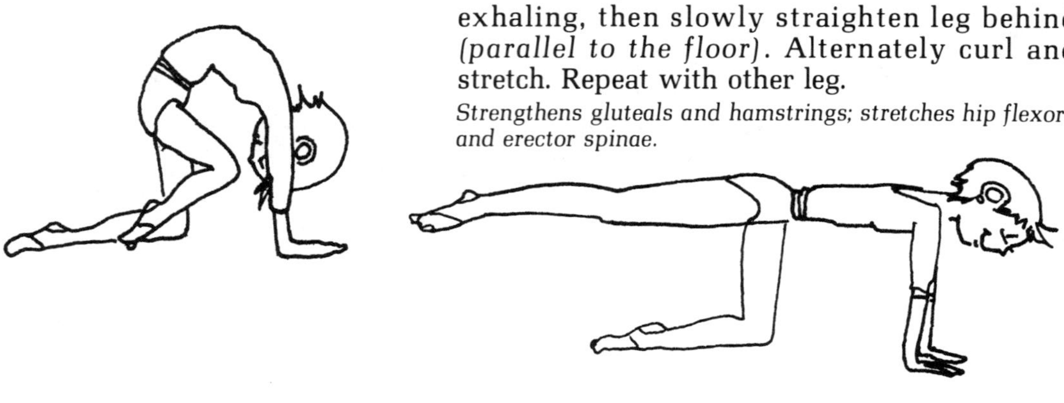

B-8 Hip Stretch

One leg straight in front, the other bent and crossed over. With arm support, twist and lean upper body toward floor on straight leg side, exhaling as you go down. Hold. Return to starting position, relax, then repeat. Same on other side.

Strengthens gluteals and hip abductors (including tensor fascia latae).

B-9 Curl-Up

See the curl-up progression on page 26. Use the technique that's right for you now.

Strengthens abdominals (including obliques if done with a twist).

Precede the curl-up with **Pelvic Tilt** (see A-8), then follow with **Apple Pick** (see A-11) to stretch the abdominals.

B-10 Side Leg Raise

Lying on one side with head resting on the forearm, raise the top leg slowly while exhaling (keeping it straight and toes pointing toward floor). Lower and repeat. Same on other side. Keep back flat throughout.

Strengthens hip abductors, and stretches adductors.

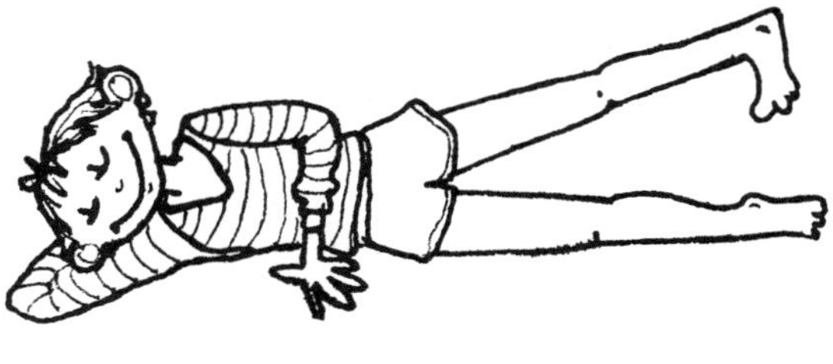

B-11 Knee Hug

Cross one leg over the other. Hug knee to chest. Hold. Relax and repeat. Change leg positions and repeat.

Stretches hip abductors.

B-12 Groin Stretch

Sitting with knees bent, soles of feet together and heels near crotch. Reach forehead toward feet while exhaling. Hold, then relax and repeat.

Stretches hip adductors and erector spinae.

B-13 Ankle Circle

Sitting, one leg crossed over the other. Grasp the foot and rotate it slowly in clockwise, then counter-clockwise, direction. Repeat with the other foot.

Stretches tibialis anterior and peroneus; lubricates ankle joint.

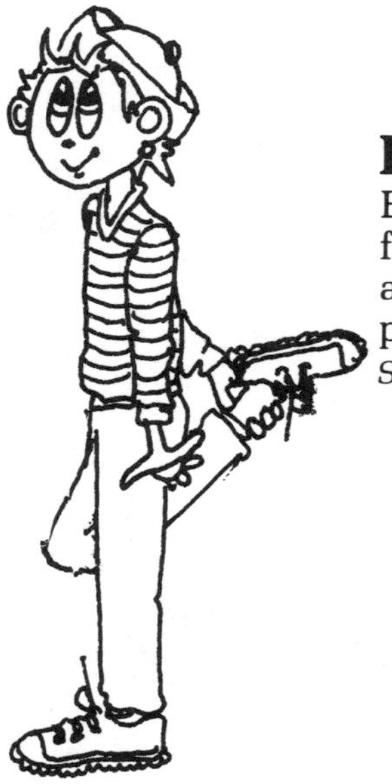

B-14 Thigh Stretch

Bend one knee, grasp the ankle behind and pull foot gently toward seat. Hold. Relax. Repeat alternately with other leg. (Maintain a proper pelvic tilt throughout.)

Stretches quads.

B-15 Soleus and 'Gastroc' Stretch

One foot in front of the other, feet pointing straight ahead and flat on the floor. Bend both legs (squatting) to stretch the soleus of rear leg. Repeat with legs further apart, back leg straight, and rocking forward over the front leg to stretch the 'gastroc' of rear leg. Use a stretch-and-hold motion. Reverse leg positions and repeat.

Stretches soleus and gastrocs.

Soleus Stretch

'Gastroc' Stretch

SERIES C

C-1 Arm Crossover

Hold arms outstretched near shoulder height. Cross them in front while exhaling, then swing them slowly behind.

deltoids, traps, rhomboids, pecs.

C-2 Neck Stretch & Overhead Side Stretch

Tip head to one side, and drop shoulder slightly then turn head to look at opposite armpit. Hold. Relax. Repeat alternately to the other side.

Stretches neck extensors and side flexors.

Reach one arm overhead and the other down the side of the leg while exhaling. Hold. Repeat alternately to the other side. Do not bounce.

Stretches erector spinae, obliques and lats.

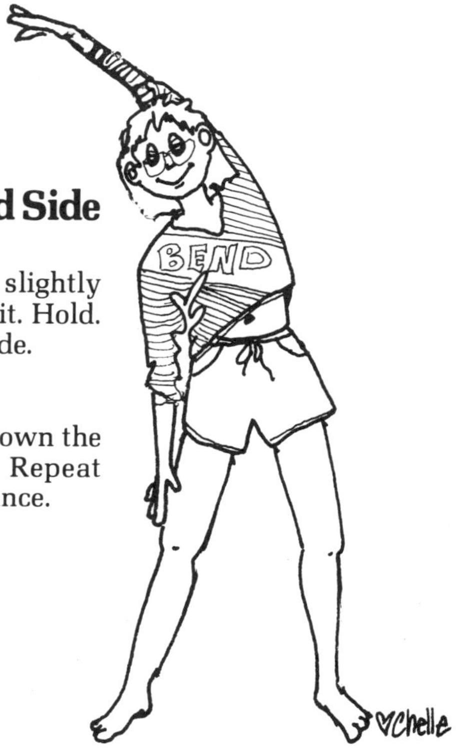

C-3 Sit Twist

Sitting with legs in front, twist to one side, reaching the arm (and looking) behind while exhaling. Hold. Repeat alternately to the other side.

Stretches abdominals (including obliques), quadratus lumborum and erector spinae.

C-4 Hamstring Stretch

On back, back flat against the floor, one leg bent with foot flat on the floor. Extend the other leg above, grasp it below the knee, and pull it gently toward chest. Hold, then relax and repeat. Same with the other leg.

Stretches hamstrings.

C-5 Wringer

On back, with arms outstretched at shoulder level and low back flat against the floor. Lift one leg to vertical then slowly drop it toward the opposite hand while exhaling. Return to the starting position, relax, and repeat alternately with the other leg.

Strengthens hip flexors, hip abductors and adductors, and hamstrings.

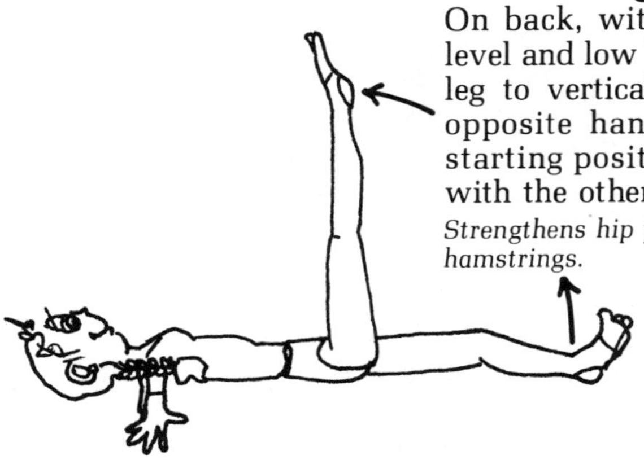

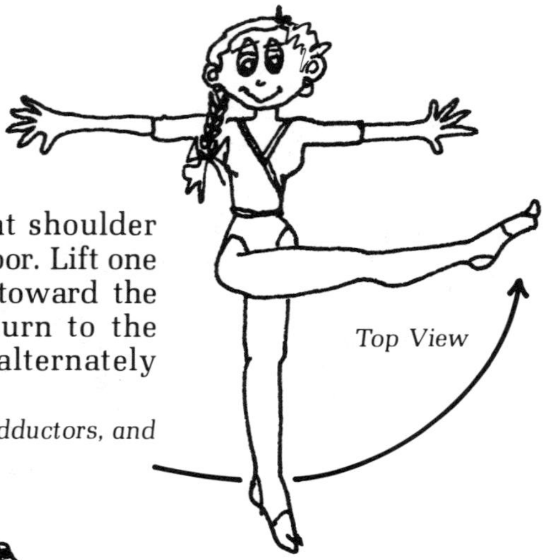

C-6 Lunge

Shift weight forward and down over bent front leg, with the rear leg kept nearly straight and the heel of the rear foot off the floor. Hold. Return to standing position and relax. Repeat alternately with the other leg. (Use a chair for balance if necessary.)

Stretches hip flexors.

C-7 Knee Toward Nose

Standing, lift one leg, grasping it behind the knee. Curl upper body forward, reaching the knee toward the nose while exhaling. Repeat alternately with the other leg.

Stretches gluteals and erector spinae.

C-8 Stretch & Tuck

On back, knees slightly bent and arms outstretched overhead, back pressed flat against floor. Curl up with the upper body and bend the legs, dragging the heels along the floor, to arrive in a seated tuck position with the hands clasped behind the knees. Exhale as you curl up. Return to starting position, relax and repeat.

Stretches erector spinae; strengthens abdominals and pecs.

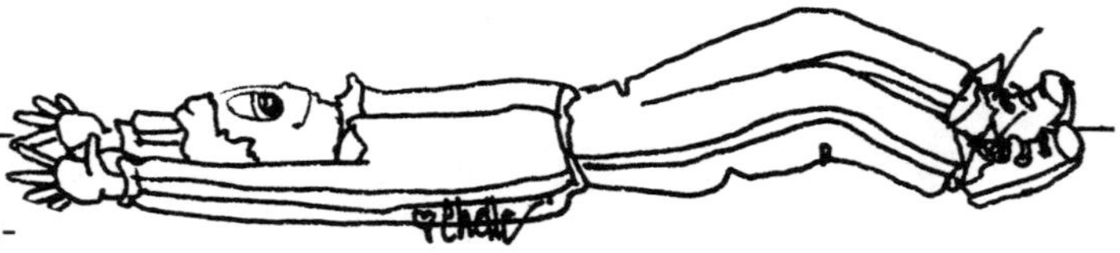

C-9 Curl-Up

Use the technique that's right for you now. See the curl-up progression on page 26 .

Strengthens abdominals (including obliques if done with a twist).

Precede the curl-up with **Pelvic Tilt** (see A-8), and follow it with **Apple Pick** (see A-11).

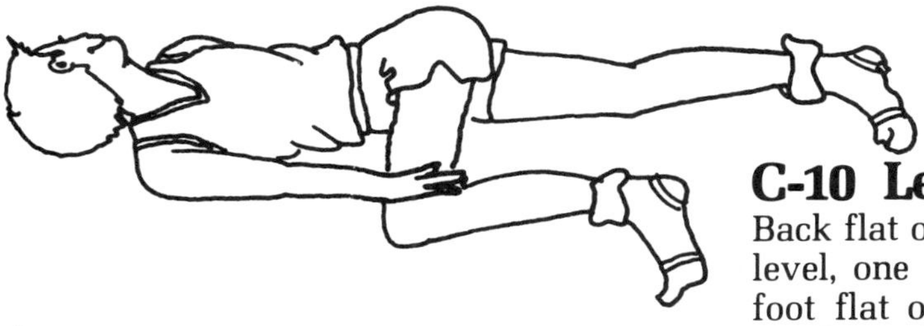

C-10 Leg Crossover

Back flat on floor, arms outstretched at shoulder level, one leg extended, the other bent with the foot flat on the floor. Roll the bent leg across, reaching the inside of the knee toward the floor. Look to the opposite side. Hold. Return to the starting position and relax. Repeat alternately to the other side.

Stretches hip abductors and abdominals.

C-11 Thigh Stretch

Lying on one side, bottom arm stretched overhead, head resting on shoulder. Bend knee of top leg, grasp the ankle behind and pull the foot gently toward seat. Hold, then relax and repeat. Roll over and repeat with other leg.

Stretches quads (and hip flexors if extended).

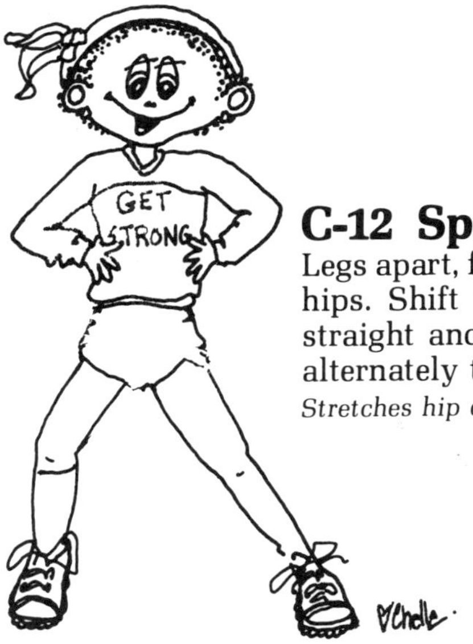

C-12 Split Stretch

Legs apart, feet pointing straight ahead, hands on hips. Shift weight over one leg, other leg kept straight and foot *flat* on the floor. Hold. Repeat alternately to other side.

Stretches hip adductors.

C-13 Side Bend

Standing with side to wall, feet flat on the floor and about 60 cm (24 in.) from the wall. Support weight with hand and drop hip toward wall, keeping upper body straight. Hold. Relax and repeat. Same on the other side.

Stretches tensor fascia latae (hip abductor).

C-14 Soleus and 'Gastroc' Stretch

Feet pointing straight ahead, one foot in front of the other, arms in front, supporting weight against a wall. Bend knees (squatting) to stretch soleus of back leg. Then, with rear leg straight behind, rock forward over front leg to stretch 'gastroc' of rear leg. (Keep heel of rear foot flat on floor.) Change leg positions and repeat.

Stretches soleus and gastrocs.

C-15 Ankle Rocker

Standing, slowly rock on outside borders of feet, from heels, to side, to toes, to other side. Knees should make a circular motion. Repeat in opposite direction.

Strengthens tibialis anterior and peroneus.

4 BACK BASICS

Back pain is one of the most common physical complaints. In fact, it has been estimated that over 80 percent of adults in Western society will develop significant back pain during their lifetime, and approximately 50 percent of these will be recurrent.

Because humans assume an upright posture, the low back must support all the upper body weight, yet remain mobile. These complex functions, if combined with the effects of poor posture, poor body mechanics, excess body weight, stress, or insufficient exercise, make the low back particularly prone to injury ... and pain. You can help your back perform well (and without pain!) by avoiding these potential problems. If you don't, upright posture becomes a burden your low back must carry.

This chapter is specifically designed to help you attain "back fitness" through appropriate exercise *and* to help you develop correct body mechanics which will protect your back during everyday activity. The information included here will be particularly useful for older adults, for those with low levels of fitness, and for individuals who have been inactive for an extended period of time.

Posture

Proper posture involves correct positioning of all levels of the spine: neck, trunk, low back, and pelvis. Test your own posture by standing with your back against the wall. Keep your heels, buttocks, and shoulder blades in contact with the wall.

Now —

- Is there a large gap (i.e. more than the thickness of your hand) between your lower back and the wall?
- Are your shoulders forward (away from the wall) *or* pushed back like a soldier?
- Is there a large gap between your neck and the wall?
- Is your chin poking forward away from the wall?

If you answered "yes" to any of the above questions, your posture can and should be improved.

Correct posture can also be *learned* using a wall. Assume the same position as for the test above.

- If you have a forward curve in your lower back, straighten it by gently tucking in your stomach, tightening your buttocks, and bending your knees slightly. These actions will help achieve an "upright" tilt of the pelvis. (See also the pelvic tilt exercise on page 22.)
- Tuck your chin in slightly and keep your head *over* your shoulders (not poked forward).
- Keep your shoulders level, slightly back, and relaxed. They should not fall in front of nor behind your hips (i.e. neither slouched nor military posture).

The exercises which follow are designed to restore and maintain back fitness by readjusting and coordinating body parts which support the back. This is accomplished by stretching tight structures, strengthening weak muscles, and developing good postural habits. As with any fitness routine, progress takes time. Don't rush or hurry your program. And practice the exercises daily.

Back Fitness Exercises

The exercises included here are *not* intended to replace a physician or physiotherapist prescribed exercise program. If you experience back pain now or you are currently receiving treatment for a back problem, consult your physician or physiotherapist regarding the use of these or any new exercises.

The rules, reminders, and cautions noted on page 8 apply equally to the exercise routine which follows.

1 Side Stretch

Reach one arm down the outside of the leg while exhaling. Hold. Repeat alternately to the other side.
quadratus lumborum, erector spinae, obliques.

2 Pelvic Tilt

Special note: Because the pelvic tilt places the low back in an optimal position for carrying the body weight and performing movement, it is a basic exercise for low back fitness. A proper pelvic tilt should be applied to all aspects of daily activities (i.e. standing, walking, lifting, carrying, etc.).
On your back, knees bent, feet flat on floor. Tighten abdominals and buttocks and press your lower back firmly against the floor while exhaling. Hold, relax, and repeat.
Stretches erector spinae.

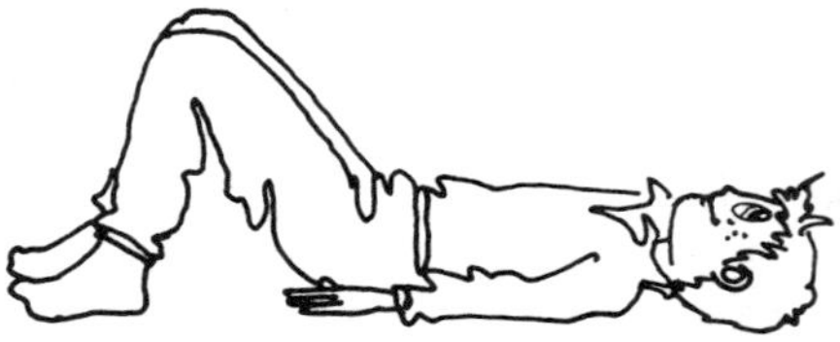

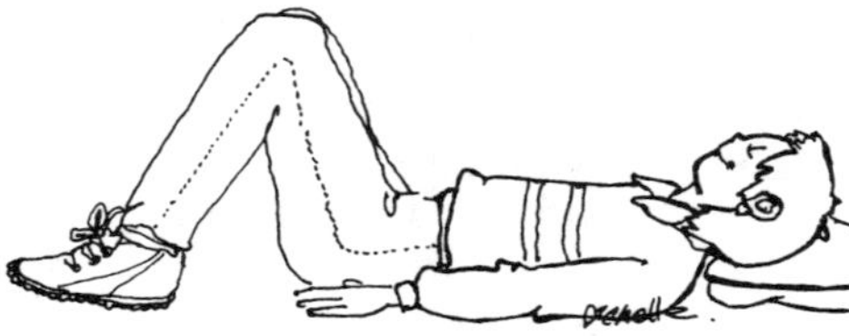

3 Hamstring Stretch

On back, back pressed flat against the floor, and hands under the seat. Bend knees to chest one at a time, and slowly lift legs — *one at a time* — to the upright position. Alternately "flap" feet up and down. (Make sure legs are bent when lifting them from or lowering them to the floor.)
Stretches hamstrings (and 'gastrocs' when feet 'flap').

4 Low Back Stretch

On back, back pressed to floor, knees bent and feet flat on the floor. Grasp one leg behind knee and pull it toward chest. Hold, relax. Repeat alternately with other leg.
Stretches gluteals and erector spinae.

5 Knee Hug

Cross one leg over the other. Hug knee to chest. Hold. Relax and repeat. Change leg positions and repeat.
Stretches hip abductors.

6 Peek-Up

On back, back flat on the floor; knees bent, feet flat on the floor. Lift head and shoulders off the floor and look toward knees while exhaling. Hold two or three seconds. Relax. Repeat.
Strengthens abdominals.

7 Lunge

Shift weight forward and down over bent front leg, with the rear leg kept nearly straight and the heel of the rear foot off the floor. Hold. Return to standing position and relax. Repeat alternately with the other leg. (Use a chair for balance if necessary.)
Stretches hip flexors.

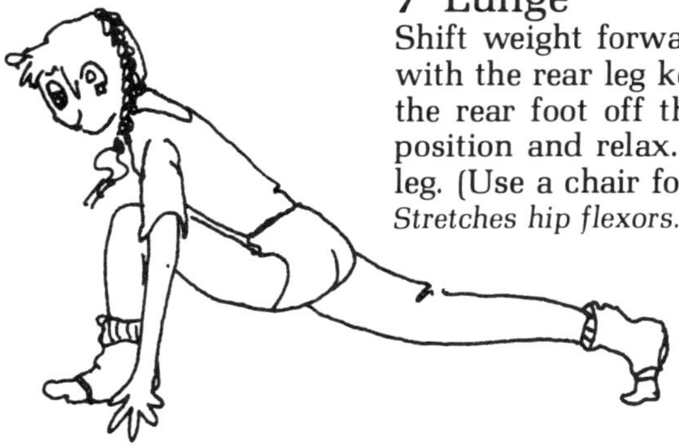

8 Butterfly

Sitting with knees bent and soles of feet together. Grasp ankles and gently press down on the inside of the knees with the elbows. Hold, relax, and repeat.
Stretches hip adductors.

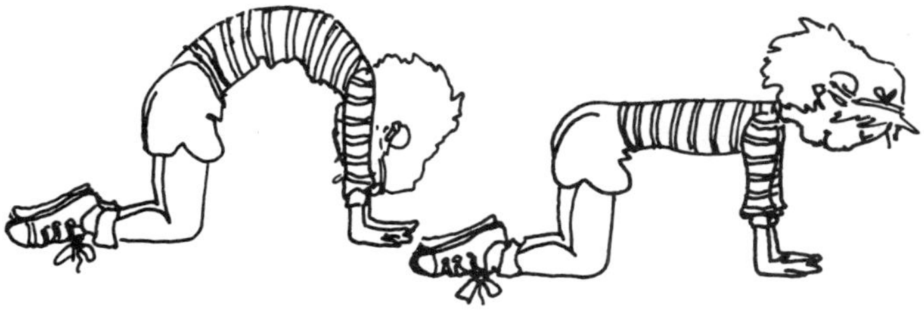

9 Mad Cat

On all fours, arch, tucking chin to chest and exhaling. Hold, then relax, returning to *flat back* position.
Strengthens abdominals and gluteals; stretches erector spinae.

Soleus Stretch *'Gastroc' Stretch*

10 Soleus and Gastroc Stretch

One foot in front of the other, feet pointing straight ahead and flat on the floor. Bend both legs (squatting) to stretch the soleus of rear leg. Repeat with legs further apart, back leg straight, and rocking forward over the front leg to stretch the 'gastroc' of rear leg. Use a stretch-and-hold motion. Reverse leg positions and repeat.
Stretches soleus and gastrocs.

Especially for the Neck and Shoulders

These exercises will be helpful if you answered "yes" to the neck or shoulders components of the posture test on page 21. All six exercises are the stretch and hold type.

1 Neck Flattening

On your back, in pelvic tilt position. Gently flatten neck to the floor by tucking your chin and slightly lifting your head.
Strengthens neck flexors, stretches extensors.

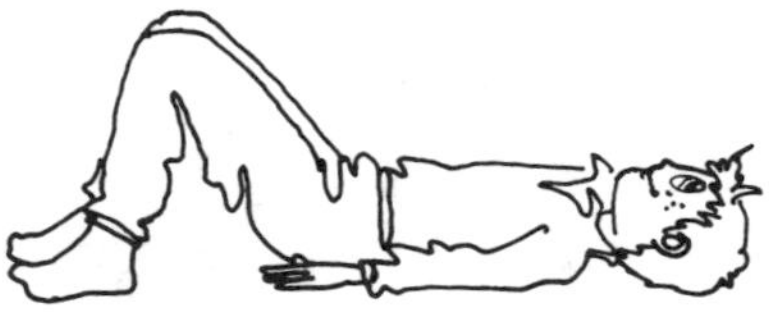

2 Shoulder Flattening

On back, in pelvic tilt position, press shoulders together and back.
Strengthens traps and rhomboids.

3 Neck Side Stretch

Standing or sitting, gently tilt head to the side, reaching ear toward the shoulder. Repeat alternately to the other side.
Stretches neck side flexors.

4 Chin to Chest
Standing or sitting, gently tuck chin to chest.
Stretches neck extensors.

5 Nose to Armpit
In a standing or sitting position, tip head to one side and drop shoulder slightly, then turn head to look at opposite armpit. Repeat alternately to the other side.
Stretches neck extensors and side flexors.

6 Pectoralis Stretch
On hands and knees, slowly "walk" your hands forward while "sitting back" on your legs.
Stretches pecs.

To Strengthen the Abdominals

The abdominal muscles work with the muscles in the back to help maintain proper posture and to help minimize the risk of developing low-back problems. Strong abdominal muscles are important for any tasks requiring bending, lifting, or carrying.

The curl-up as shown below is an effective exercise for developing and maintaining abdominal strength. Experiment with the different arm positions and movements shown and begin with the one that allows you to do two sets of 8 to 10 repetitions *without* straining or overexertion. Use that method when you do your curl-ups in Series A, B, and C. When one method becomes too easy, move on to the next more difficult one. (They get *gradually* more difficult as you move from *Curl-up with elbow roll* through to *Curl-up with a twist.)*

Avoid anchoring the feet down (this reduces the demands on the abdominals), and **avoid** doing the curl-up with legs straight (this puts undue pressure on the lower back).

Curl-up progression. Regardless of the method you use, begin on your back with your knees bent and feet flat on the floor. *Curl up* to a sitting position tilting your pelvis and pressing your lower back to the floor as you begin the movement. Exhale as you curl up.

Curl-up with elbow roll...
Lean slightly to one side and push off with your elbow and hand as you curl up. Curl down the same way. Lean alternately to the other side.

Grasp Thighs...

Hands at Sides...

Arms across Chest...

Hands at Ears...
(*not* behind head)

Twist
Twist gently *at the end of the curl-up* reaching elbow toward the opposite knee. Twist alternately the other way. (Keep hands at ears throughout movement.)

Everyday Mechanics

Correct posture and strong, supple muscles combined with proper body mechanics will help you ensure healthy and efficient back function. Following are some basic guidelines for everyday activities.

Sitting

- Use a footrest to keep your knees higher than your hips.
- Slide your hips slightly forward so your upper back is against the chair support — avoid arching your neck or back.
- When writing or reading at a desk, bend forward from your hips, not your neck.

Standing

- Use a footrest (a box, stool, or seat of a chair) to raise one foot higher than the other.
- Keep your knees slightly bent whenever leaning forward.
- Practice pelvic tilt.

Walking

- Remind yourself frequently of the elements of good posture — pelvic tilt, chin in, shoulders down and back.

Lifting and Carrying

- Pelvic tilt first, then…
- Keep a stable base — feet apart and knees bent.
- Lift with your legs — don't make your back do all the work.
- Squat with one foot in front of the other when picking up objects or putting them down.
- Keep objects close to your body.
- Hold objects at waist height when carrying; avoid lifting objects overhead.

Resting and Sleeping

- Keep knees bent and pelvis tilted when lying on your back; support your knees with a pillow or rolled blanket.
- Keep your knees bent to flatten your lower back when lying on your side — bend bottom knee slightly more than the other.

Finally

- **Avoid** twisting at your waist while carrying objects. Move your feet instead.

Face on …

don't twist

- **Avoid** bending forward at the waist *for long periods of time* (eg. raking leaves, vacuuming, etc.) Remember to bend your knees when you bend forward.
- **Avoid** arching your back or bending backward.

5 PREPARING FOR SPORTS AND RECREATIONAL ACTIVITIES

Different sports place different demands on the body. While a warmup stretching routine should include all major muscle groups, you may want to spend additional time on the muscles which are used most in your particular activity.

The chart below notes a variety of sports and indicates the major muscles involved in them. The activities are categorized by their dominant motions and muscle actions, and the appropriate exercises (from Series A, B, or C) for stretching the specific muscles are listed for easy reference.

Remember that a good level of flexibility and a proper warmup each session will not only help you perform better but also prevent unnecessary injury. And as a general rule, the more intense the activity, the longer the warmup stretching period should be.

Stretching for Sports and Recreational Activities

Activity	Major Muscles	Exercises
Baseball, Fastball	deltoids, triceps, teres, quads, hamstrings, soleus, gastrocs	A-1, C-1, C-2, A-4, A-12, C-14
Golf	erector spinae, lats, obliques	B-3, B-7
Basketball, Volleyball, Alpine Skiing, Figure Skating	quadriceps, gluteals, gastrocs, hamstrings, soleus, hip adductors	C-11, C-14, A-4, C-12
Racquet Sports (Tennis, Racquetball, Squash, Badminton), Handball	deltoids, pectoralis, lats, hip adductors, gluteals, gastrocs, soleus, quads, tensor fascia latae	A-1, C-1, C-2, C-12, C-14, A-5, B-10
Cycling	quads, gluteals, gastrocs, soleus, erector spinae	B-14 or C-11, C-14, A-5 or C-7
Speed Skating	quads, gastrocs, hip adductors, tensor fascia latae, gluteals, erector spinae	B-14 or C-11, C-14, A-13, B-8, A-5 or C-7
Canoeing, Rowing, Kayaking	obliques, deltoids, lats, teres, triceps, biceps, erector spinae, quads, hip flexors	C-1, C-2, C-3, B-1, A-5, B-14, C-6
Field Hockey, Ice Hockey, Soccer	obliques, deltoids, erector spinae, quads, gluteals, tensor fascia latae, hip adductors	B-1, C-2, B-3, C-13, B-14, C-12
Football, Rugby	quads, gluteals, tensor fascia latae, hip adductors, gastrocs, soleus, deltoids, trapezius	B-14 or C-11, C-13, C-12, C-14, B-1
Running	hamstrings, gastrocs, soleus, hip flexors	A-4 or C-4, C-14, C-6
Cross-country Skiing	quads, hip flexors, gastrocs, soleus, gluteals	C-11 or B-14, C-6, C-12, C-14, A-5
Swimming	deltoids, pecs, lats, triceps, trapezius	B-1, C-1, B-3

ABOUT THE CONTRIBUTORS

GORDON STEWART does consulting, writing, and editing in the areas of health, fitness, and sports conditioning. He has done work for a variety of organizations including Sun Life Assurance Company, MacMillan Bloedel, Fitness Canada, and the Arizona Public Service Company. Among his publications are *Every Body's Fitness Book* and *Effective Fitness Management.* He holds an M.Sc. in kinesiology from Simon Fraser University.

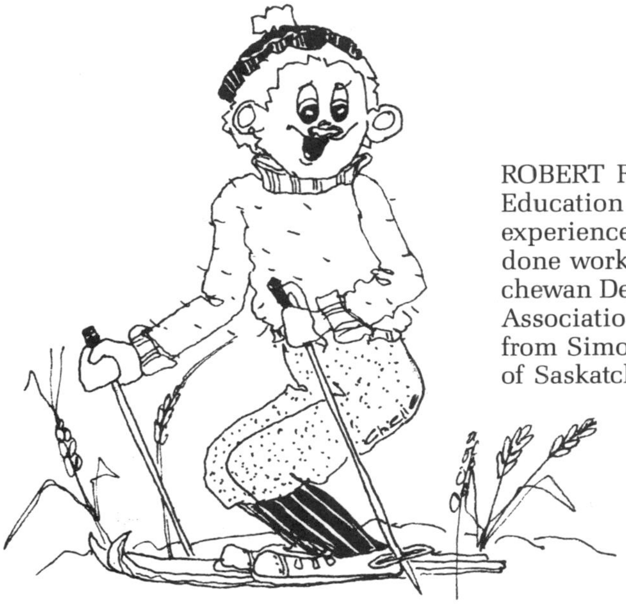

ROBERT FAULKNER is a professor in the College of Physical Education at the University of Saskatchewan. He has extensive experience in adult fitness/exercise management, and he has done work for a number of organizations including the Saskatchewan Department of Health, the Canadian Parks & Recreation Association, and ParticipAction. He has an M.Sc. in kinesiology from Simon Fraser University and a Ph.D. from the University of Saskatchewan.

CHELLE (Richelle Bradshaw) is a dance and fitness instructor, and freelance artist — specializing in cartooning. Many of her clients are fitness and recreation organizations, but she has also done work for Eaton's, Woodward's, and MacDonald's.

ABOUT THE EXERCISES

The exercises included here are conservative and meant to be safe and beneficial. Some good exercises have been purposely omitted, experience showing that they could not be safely and effectively communicated by words and picture alone.

Many of the exercises are adapted from those used in Dr. William Ross' exercise management course at Simon Fraser University and the adult fitness program founded by Dr. Fred Kasch and Dr. John Boyer at San Diego State University.

ADDITIONAL COPIES — ORDERING INFORMATION

Ordering Policy. Orders from individuals and all orders under $50.00 must be prepaid. Postage / handling charges: For orders less than $100 add 10%. For orders $100 or more add 7%.

For **quantity orders** (10 or more copies of any one title) a 25 percent discount is available. Further discounts are available for orders of 100 or more copies of any title. Write for details.

BEND & STRETCH Suppleness and Strength Exercises

ISBN 0-920846-03-3, paperback, $4.95 (10 or more copies – $3.70 each).

FIT ALL OVER A Catalogue of Exercises

by Shirley Main, Gordon W. Stewart, and Richelle Bradshaw

ISBN 0-920846-04-1, paperback, $8.95 (10 or more copies – $6.70 each).

FIT ALL OVER provides an extensive selection of exercises to improve strength, suppleness, stamina, mobility, coordination, and balance. Basic exercises, variations, and exercise sequences are provided for the hands and wrists, shoulders and arms, chest and back, abdominals and hips, and legs and feet. Whole body exercises, aerobic activities, and sample workout routines are also included.

GET STRONG A Sensible Guide to Strength Training for Fitness and Sports

by Brian Cook and Gordon W. Stewart

ISBN 0-920846-01-7, paperback, $6.95 (10 or more copies – $5.20 each).

GET STRONG includes information on weight training facts and fallacies, principles for program design, core training routines and an exercise catalogue, injury prevention, women and weights, purchasing home equipment, designing programs for children, and guidelines for training for sports and recreational activities.

EVERY BODY'S FITNESS BOOK A Simple, Safe, and Sane Approach to Personal Fitness

by Gordon W. Stewart, M.Sc.

ISBN 0-920846-02-5, trade paperback, 240 pages, $12.95 (10 or more copies – $9.70 each).

EVERY BODY'S FITNESS BOOK is based on the premise that getting fit *can* and *should* be painless, positive, and personally meaningful. In a friendly and informal style, it answers the fitness questions most commonly asked and simplifies the issues most often misunderstood. It provides people of all sizes, shapes, ages, interests and aptitudes with the information necessary to create an individualized fitness plan that works.

These publications are sold through some recreation centres, employee fitness programs, and other health and fitness clubs and organizations. If they are unavailable in your area, please order directly from the publisher. Fitness programs wishing to make them available to their participants are advised to obtain quantity orders from the publisher for resale to participants at the retail price.

3S® Fitness Group Ltd.
Box 5520, Station B
Victoria, British Columbia
V8R 6S4
Phone (604) 598-1426
Fax 598-1748

Exercise wall-charts adapted from *BEND & STRETCH, FIT ALL OVER* and *GET STRONG* are also available. The charts are designed for use in exercise rooms, gymnasiums, weight training areas, dance studios, schools, pools, recreation centers, fitness and sports clubs, and apartment and condominium exercise areas. They provide proper programming and thus aid in injury prevention in situations where fully supervised activity is not always possible or appropriate. Six charts are available in all. Write for details and costs.

A copy of our catalogue which describes all our products and publications is available FREE on request.